FRICTION WELDING

The tendency of two metals to seize together when relative movement, heat and pressure are applied has been recognised as an engineering embarrassment for as long as machines have been running. Patents were taken out many years ago to cover the commercial use of the principle as a means of metal joining. Only in the last decade, however, has friction welding as such become an established method of metal joining.

This monograph describes in brief the fundamental aspects of friction theory, but the bulk of the text is concerned with how the technique is being used in an ever increasing field of application ranging from high output rate automobile components to the manufacture of very large joints. There is little doubt as to the commercial viability of friction welding and this is reflected in the numerous examples quoted where economies have been effected.

Other titles in this series

All published by Mills & Boon Limited

M & B Monograph   ME/13

General Editor: J Gordon Cook, PhD, FRIC

Friction Welding

M J Fletcher, Ph D

Advisor on Materials Selection and Joining

Distributed in the United States by
CRANE, RUSSAK & COMPANY, INC.
52 Vanderbilt Avenue
New York, New York 10017

First published in Great Britain 1972
by Mills & Boon Limited, 17–19 Foley Street,
London, W1A 1DR

© Mills & Boon Limited 1972

ISBN 0 263.05108.0

Made and printed in Great Britain
by Butler & Tanner Limited, Frome and London

CONTENTS

1. Introduction

The use of frictional heat for joining is not a recent concept. When metal is being turned on a lathe it will sometimes weld itself to the tail stock; similarly, swarf may become welded to the cutting tool. In reciprocating engines the problem of seizure due to lubrication failure has been with us for many years.

The first patent covering the use of frictional heat in welding was taken out in Britain in the early 1940s. The first practical application of the technique, about a decade later, has been credited to Chudikov in the U.S.S.R.

These early practical applications were followed by a comprehensive investigation of friction welding, but it was not until 1960 that work was initiated commercially in the United Kingdom. Friction welding machinery has subsequently developed from what were essentially converted lathes—adequate for basic test work and evaluation—to sophisticated and automated units ranging in size from micro machines to 80,000 kg monsters. Commercial machines capable of joining cross-sectional areas from 0·2 to 600 mm diameter are now available. Many are operating in environments as demanding as flow line production in the automobile industry where repeatability of joining and high throughput rates are of paramount importance.

Despite the tremendous impact that friction welding has had on metal joining generally there is relatively little published work on the subject. This is probably a reflection in the first instance of the rate of change in machine design. Technological development in friction welding machinery applications has progressed at such a pace that it has been difficult to maintain a correspondingly high output of published work. Furthermore, our understanding of the theoretical aspects of friction welding has not kept pace with its practical development.

2. Fundamental Aspects

Welding by friction is achieved by rubbing two surfaces together such that heat is generated, followed by the application of a force normal to the plane of the joint.

Friction has been the subject of intensive research during the past 30 years. Yet there is still no theory capable of explaining and predicting the behaviour of materials subjected to frictional forces. As a result, there are no established guidelines to assist the user of the friction welding process in selecting optimum working parameters for any given joint.

Superficially, the technique of friction welding is simple. Fundamentally, it is a complex phenomenon in which many inter-dependent mechanical and metallurgical processes are involved. Many of these processes have been examined individually, but there has been no overall co-ordination of the variables involved. Considerable work remains to be done before a thorough understanding of the various mechanisms can be presented.

Friction welding is an extension of the wide range of solid-state bonding processes now in commercial use for joining. It is a member of the family of techniques which includes flash welding and the traditional forge welding, where heat is generated at the joint interface and pressure is applied to force joint components into intimate contact.

The nature of friction welding may be illustrated by examining the various mechanisms which contribute to the formation of the joint. The technique is represented diagrammatically in its simplest form in Fig. 2.1. From this it will be observed that two principal machine variables are involved; relative speed of rotation and normal load.

When two practically flat surfaces are brought into con-

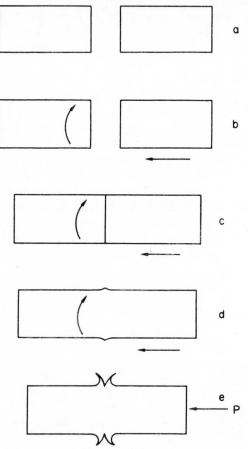

Fig. 2.1. Schematic diagram showing basic features of the friction welding technique. The two originally static and clamped parts are shown in (a). The left hand part is rotated (b) and then the two are brought together with light pressure on the right hand component. This causes frictional heating at the interface and a limited upset (d). The forge pressure P is applied (e) and relative rotation stopped to make the final joint.

tact, only discrete projections on the abutting faces actually touch one another (Fig. 2.2). This is because even a highly polished surface has microscopic irregularities, projections up to 10^{-9} mm being measurable. The actual

9

physical contact area (A_a) is thus less than the geometric area (A_g). The ratio $A_a : A_g$ depends upon the machined finish of the surfaces. To establish an optimum joint the entire surface areas must contact and bond, and it is obviously necessary to cause the ratio A_a to A_g to become unity. This is realised in friction welding by forcing the joint surfaces together, thus causing severe elastic and plastic deformation on and near the joint. Relative motion of the surfaces which also occurs in friction welding serves two purposes. Firstly, heat is generated, causing a decrease in the strength of the material; in consequence, the load required to effect intimate overall contact is lowered. Secondly, repeated relative motion causes destruction of some projections and re-distribution of the applied load amongst remaining ones. Under these circumstances the entire actual contact area A_a undergoes uniform wear.

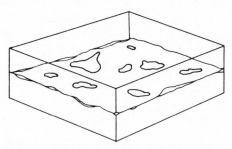

Fig. 2.2. Two surfaces placed together depicting the actual areas in contact.

Actual contact area is thus a function of (1) the surface condition of the contacting materials (i.e. the smoothness), (2) the applied load and (3) the strength of the materials under consideration (i.e. their resistance to deformation). Expressed simply:

$$A_a = k_1 P + k_2 \qquad 1(a)$$

where P = applied force
k_1 = constant proportional to the strength of the material
k_2 = constant dependent upon the surface of the material.

FRICTION FORCE

Two hypotheses have been advanced to explain the nature of the friction force, i.e. that force which resists relative movement between two bodies in contact. The mechanical hypothesis suggests that macro physical inter-locking and elastic and plastic deformation are responsible for the resistance to relative movement (Fig. 2.3).[1] Contacting and interlocking projections must be elastically or plastically deformed in order that relative displacement of the contacting surfaces can take place. The net energy required to cause deformation over the entire surface is equal to the friction force. A more recent hypothesis based upon molecular considerations has been advanced. In this, the attraction of bodies in molecular proximity to one another is considered.

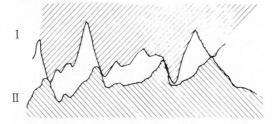

Fig. 2.3. The mechanical hypothesis—purely mechanical interlocking of projections of two continuous surfaces I and II.

It is unlikely that either of these two hypotheses could individually explain the mechanism of friction. A combination of mechanical and molecular forces may go some way towards elucidating the basic phenomena, but much research remains to be carried out before the overall mechanisms can be fully comprehended.

FRICTION COEFFICIENT

Work undertaken in the seventeenth century by Amonton established that the friction force was a function only of the applied pressure, i.e.:

$$F = \alpha P \qquad \qquad 1(b)$$

11

where α = constant which Amonton described as the "coefficient of friction"

F = friction force

These early experiments were not substantiated by subsequent work undertaken by Coulomb in the eighteenth century. Coulomb found that the coefficient of friction diminished with increasing pressure. Recent investigators have established that the coefficient of friction is dependent upon three principal factors; pressure, relative velocity and material properties. Independent research directed at evaluating a relationship between α and one of the three inter-dependent factors has led to the publication of apparently contradictory results. These have confused rather than clarified the state of knowledge.

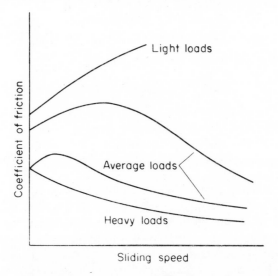

Fig. 2.4. Effect of speed and normal loading on friction. [2]

It is now widely held that all three variables—pressure, speed and material properties—contribute significantly to the friction coefficient and therefore to the mechanism of the friction welding. Evidence to support this belief is shown in Figs. 2.4[2] and 2.5.[3] Fig. 2.4 demonstrates the influence of speed and pressure. Fig. 2.5 shows how different materials dictate welding machine torque, i.e. the frictional force required to sustain constant load and speed.

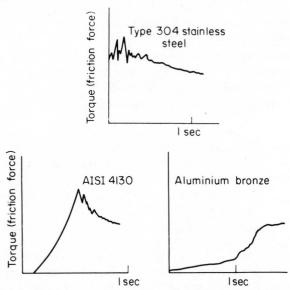

Fig. 2.5. Friction characteristics for three commercial alloys under conditions of continuous heating from ambient temperature and constant normal load. [3]

Additional complications arise when the two materials subjected to friction phenomena are dissimilar, particularly if one material is significantly softer than the other. This was recognised as far back as 1954 when Bowden and Tabor[4] proposed a mechanism which takes into account not only the mechanical concept illustrated in Fig. 2.3 but also the fact that a hard material will plough into a softer material. They expressed the coefficient of friction as follows:

$$\alpha = \frac{S}{P} = \frac{1}{(aK^{-2} - a)^{\frac{1}{2}}} \qquad 1(c)$$

where S = interfacial shear strength

a = constant whose value is dependent upon the softer material; usually the figure is about 9

13

> K = proportionality factor which is a function of surface cleanliness in the metallurgical sense. This will change during the course of the friction welding cycle as the surface metallurgy of the joint changes.

A further attempt to quantify the phenomena of friction has been made by Goddard and Wilman[5]. The equation, developed by them and given below, separates the shearing and ploughing mechanisms;

$$\alpha = k_1 S_1 P^{(2-n)/n} + k_2 S_2 P^{(3-n)/n} \qquad 1(d)$$

S_1 = interfacial shear strength
S_2 = dynamic shear strength of the softer material
P = applied force
n = measure of material hardness lying between 2 (hard) and 2·5 (soft)
k_1 and k_2 are constants

Note that in neither of the equations 1(c) and 1(d) is there a term which accounts for influence of relative velocity. Kragel'skiy and Vinogradova[2] produced a relationship (1(e)) which takes velocity into consideration, but is appropriate only for constant loads;

$$\alpha = (A + Bu)e^{-Cu} + D \qquad 2(e)$$

where V = velocity
A, B, C and D are coefficients

This equation has been used to derive points for Fig. 2.4.

Despite numerous attempts to analyse the mechanism of friction welding, therefore, no single treatment has yet been developed which completely accommodates all the variables known to be effective in the friction phenomenon. The theories are helpful in establishing certain fundamental aspects, but can be disregarded from the practical point of view. Working parameters must be evaluated from previous experience. A substantial volume of documented information on materials, optimum speeds and loads is available from friction welding machine manufacturers and from existing users.

TEMPERATURE

Since material properties are temperature dependent, the overall friction mechanism as represented by equations 1(a) to 1(d) will be markedly influenced by temperature changes.

Additionally, heat generated during the friction part of the cycle also influences the forging pressure, i.e. the final force applied normal to the joint necessary to effect welding. If the temperature is low, a higher forging pressure will be needed than if the temperature is high. This is simply because materials generally are more easily deformed at higher temperatures.

Furthermore, the temperature of the interface immediately prior to the welding operation has a very significant effect upon the final physical and metallurgical properties of the joint.

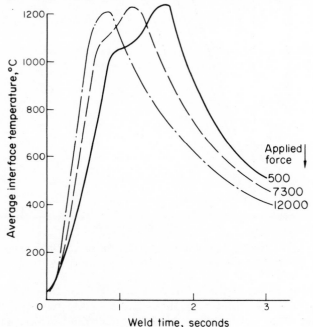

Fig. 2.6. Effect of varying welding pressure on interface temperature. [3]

Clearly, temperature at the joint interface is most important. One of the more interesting facts to emerge from contemporary research is that the temperature at the interface is never sufficiently high to allow molten metal to be produced.[6] This is the case when pure identical metals are being joined or when low-melting-point eutectic alloys form as a result of the welding of dissimilar materials. Furthermore, the maximum temperature is essentially independent of the applied force (Fig. 2.6)[3] and of the holding time under frictional conditions.

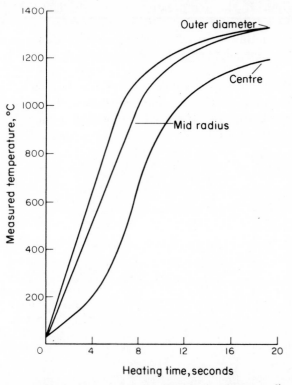

Fig. 2.7. Radial temperature distribution in 20 mm diameter friction welded low carbon steel 1 mm from interface.[7]

Examination of temperature distribution at the interface has revealed that a gradient exists in the radial direction (Fig. 2.7) even in prolonged welding cycles. This gradient will clearly depend upon the thermal conductivity of the material or materials being joined, and the temperature differential between inner and outer portions of a joint will increase as the diameter increases. Such an observation could explain why, under certain conditions, only welding in the peripheral regions occurs leaving the central part unwelded.

3. Process Variables

In the previous chapter the more fundamental aspects of friction theory were outlined. We can now examine in practical terms how the variable machine parameters—rotational speed, applied pressure and time—influence the way in which a weld is formed.

For any welding process there is an optimum combination

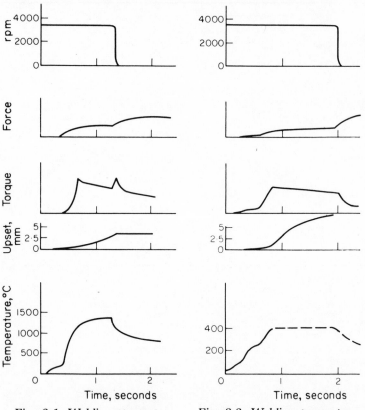

Fig. 3.1. Welding parameter relationship for iron–nickel (8 mm diameter solid bar).[8]

Fig. 3.2. Welding parameter relationships for aluminium–titanium (8 mm diameter solid bar).[8]

of parameters with which a weld can be made. In some techniques, e.g. electron beam welding, satisfactory joints can only be produced by maintaining close control over these set conditions. Even minor variations can lead to defective welds. With friction welding the tolerance band for deviation from the optimum set conditions is generally quite wide. Entirely acceptable welds can be produced even if relatively large changes in working parameters occur. This is not to say that the process is capable of absorbing slipshod work, but it does permit a little extra freedom in the machine programming. It also offers the

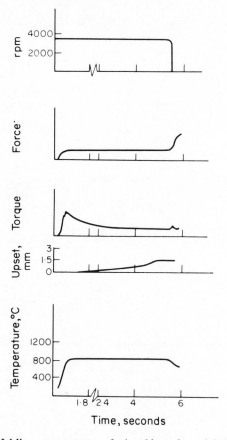

Fig. 3.3 Welding parameter relationships for nickel–titanium (8 mm diameter solid bar).[8]

opportunity for fully automatic operation without the need for highly sophisticated and expensive control of working conditions.

Since material properties exert such a significant influence upon the overall mechanism of friction welding it is not possible to generalise concerning parameters which should be used in a given practical situation. This is illustrated graphically in Figs. 3.1–3.5[8] which show temperature/upset and applied force variation for five combinations of materials.

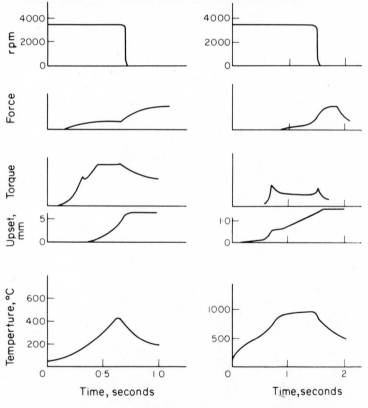

Fig. 3.4. Welding parameter relationships for aluminium–nickel (8 mm diameter solid bar).[8]

Fig. 3.5. Welding parameter relationships for nickel–titanium (8 mm outside diameter tubing).[8]

For practical purposes the friction welding cycle can be divided conveniently into two stages: (1) heating and (2) forging.

(1) HEATING

In the heating phase, the two members are allowed to rub against one another under moderate loads to generate heat. During this stage soft metal is produced at the interface and extruded axially under the effect of the applied load. A maximum temperature is attained at the junction which is below the melting point of the lowest-melting-point metal or alloy (see Chapter 2). Various combinations of speed and applied load can be used to create this condition, different materials demanding different parameters to effect the best results. For example, it is widely believed that copper, molydenum and tungsten demand much higher speeds than ferrous alloys, aluminium and lead. The actual velocity will obviously be a function of bar diameter or—in the case of tube—wall thickness, since heavier sections will present larger heat sinks. This is particularly important in the case of copper, which has a high thermal conductivity.

For a given material, relatively low speeds tend to have the effect of increasing the demand on machine torque; this creates problems in holding the work piece. In addition, heating is generally less uniform. High velocities give rise to larger heat-affected zones.

Higher pressures mean greater torques and increased power requirements. For example, doubling the pressure on mild steel joints demands a 50% increase in drive power to compensate for additional torque developed. Steeper temperature gradients at the joint interface are also generally obtained when using higher pressures.

The higher rates of heat input associated with high pressures may be desirable from a commercial point of view, but there is a tendency to produce a non-parallel heat-affected zone and non-uniform heating. This can lead to

21

the production of defective welds or, at best, undesirable metallurgical structures in many materials.

The pressure used will depend to a large extent upon the mechanical properties of the materials being joined. Materials such as tool steels retain their strength at elevated temperatures. With mild steel, on the other hand, strength falls off rapidly as the temperature increases. Tool steels will thus require higher pressures than mild steel to effect the necessary plastic deformation at the joint. Fig. 3.6 illustrates the influence of temperature on a range of materials in common use.

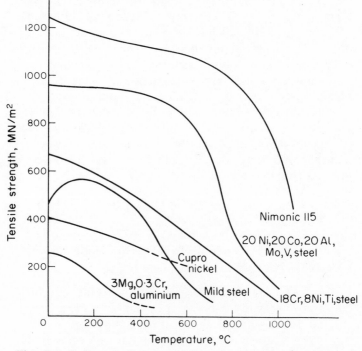

Fig. 3.6. *Influence of temperature on the strength of a range of materials in common use.*

(2) FORGING

When the desired temperature has been attained at the

joint interface the second phase of the operation—the forging process—takes place. Here, the driving power is removed and the frictional resistance is allowed to halt the relative movements of the components. Additional braking may also be used (see Chapter 4). As the drive is cut off the applied pressure is generally increased. This new force is called the "forging pressure".

The forging phase is responsible for the final gross plastic deformation inherently associated with friction welding— commonly referred to as the "flash" or "collar". It is necessary to effect optimum joint properties by bringing the two mating surfaces into intimate contact and axially extruding any contaminants which might still be present at the interface.

It has been indicated previously that the applied pressure, relative speed and duration of each of the two stages of friction welding depend largely upon the properties of the material or materials being joined together. They are also dependent upon the type of equipment used for the welding operation. Although the fundamental aspects of friction phenomena are common to all types of machine, working parameters necessary to produce a satisfactory joint can vary widely.

Equipment can be divided broadly into two basic forms. The first to be developed and the one in most common use at the present time is *continuous drive welding*. A motor drive is employed throughout the cycle which can thus be maintained for an indefinite period. The second and more recent development is *inertia* or *stored energy welding*. A heavy flywheel is activated, the drive disengaged and the inertial energy in the flywheel used to provide power for the welding cycle. Table 3(a) describes the basic features of the two forms of machine. Fig. 3.7[9] illustrates how the power demand/time characteristics compare.

One effect of having a fixed quantity of energy stored in a flywheel is that, given a rate of dissipation of power into a joint, the time cycle is then fixed. The practical

Table 3(a). Comparison of the basic features of continuous drive and inertia welding

CONTINUOUS DRIVE WELDING	INERTIA WELDING
Total energy defined by setting time. Burn-off and therefore length can be altered to accommodate variations in parts	Energy limited by power input into flywheel. Once in motion this cannot be varied
Larger drive units needed since friction cycle must be powered by motor. Smaller sections can be joined on portable equipment	Drive units light since only have to impart inertia to flywheel. Heavier plant necessary to react to high torque forces. Weld section limited by mechanics of equipment and not be drive motors.
Relatively long cycle times	Cycle times can be less than 1 second
Energy input 0·1–0·5 kpm/sec³	Energy input 0·3–2 kpm/sec³

outcome is that deviations from previously selected parameters cannot be accommodated. This is no disadvantage when initially assessing the optimum machine settings, since variations in total flywheel energy could be made to achieve a desired result. Once the machine is on the shop floor and established in a production line, however, limits are set. If a tolerance on finished part length should be mandatory, therefore, the components must be machined to achieve this tolerance. With continuous drive welding, wide deviations in machining can be accommodated by maintaining the rubbing and extruding action of the heating phase until the parts are of the correct length. This can be effected by the use of simple limit switches.

The power/time characteristics (Fig. 3.7) for inertia and continuous drive welding can be markedly different. Although the total power demands are not widely different, the rate of dissipation is relatively high for inertia welding and is reflected in small weld-time cycles. This is attractive in production applications, but in some materials can give rise to steep temperature gradients and associated metallurgical conditions which may be undesirable.

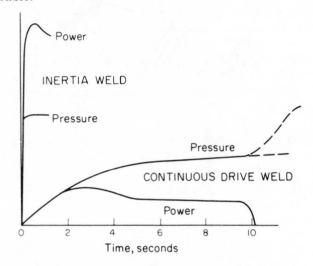

Fig. 3.7. Variations in power and pressure requirements between continuous drive and inertia welding.[9]

Although heavier plant is required for inertia welding to overcome the short-duration high-torque forces involved (see Fig. 3.7), smaller drive units can be used. Motors are only required to bring the flywheel to a pre-determined speed. They can then be disengaged. Joint cross-section in the case of inertia welding is thus predominantly a function of mechanical strength of the machine. This is in contrast to the continuous drive method, where power available from the drive mechanism is the over-riding factor.

4. Equipment

Once the principle of applying frictional heat to the welding of metals had been recognised there was fairly widespread interest throughout manufacturing industry. In the 1950s early experiments were undertaken using conventional lathes. It was quickly observed that this was not a satisfactory arrangement, since metal-turning equipment is incapable of satisfying the requirements of high speeds and high applied pressures necessary to effect suitable joints by friction welding. The additional need for extremely high torques during the forging operation placed demands on some machines which caused catastrophic failure of the drive and work-holding mechanisms.

Conversion of conventional lathes for friction welding purposes has met with some success. Available information indicates, however, that the cost of conversion is often significantly greater than the cost of a standard friction welding machine.

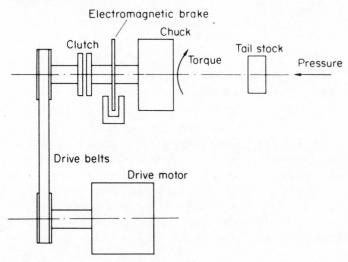

Fig. 4.1. Schematic illustration of continuous drive welding machine.

Examination of the basic mechanical requirements for friction welding machinery reveals that special attention must be made to fundamental design. The schematic drawing of a continuous drive welder (Fig. 4.1) illustrates the major demands placed upon a machine.

High applied pressures between tail stock and chuck, and torques developed at the joint and transferred into the drive and tail-stock assemblies, dictate that rigid mechanical support is mandatory. A typical value for heating load is 40 N/mm²; a typical value for forging is 85 N/mm². The mechanical support usually takes the form of two heavy tie bars. These also permit adjustment of the distance between chuck and clamp by allowing the latter to be unlocked and to slide along the bars. Such a facility is necessary where different assemblies for welding may vary in length. Fig. 4.2 shows a typical production machine which incorporates the tie bar principle.

Fig. 4.2. Tie-bar principle friction welding equipment. (Courtesy Gatwick Electrode Ltd.)

An alternative to the tie bar technique is to employ a heavy base which resists bending and is thus capable of accommodating the high applied loads on specimens. This is a more recent innovation than the tie bar method and it allows greater operator and maintenance access. A machine which employs this type of mechanical restraint is illustrated in Fig. 4.3.

Fig. 4.3. Heavy base type machine. (Courtesy John Thompson Ltd.)

Apart from the more obvious ancillary equipment requirements such as a drive motor, chuck and clamp, some means of stopping the rotational motion when the weld has been made may be necessary. With inertia welding a separate mechanical system is not used, since the momentum in the fly-wheel is calculated to deliver just sufficient energy to the work pieces to ensure a satisfactory joint. Continuous drive machines must, however, be stopped when the weld has been formed. This can be achieved by disconnecting the drive mechanism from the chuck using a clutch and applying a mechanical brake, by electromagnetic braking or by braking of the drive motor itself. Most machine manufacturers tend to favour the clutch mechanism in which the entire drive unit is disengaged from the work assembly and the latter braked independently.

In the United Kingdom continuous-drive friction welding has seen much greater application than inertia welding, the latter being more favoured in the United States and in some continental countries. Overall there is little to choose between them. Advantages of one are in general cancelled out by advantages in the other all the way down the line from manufacture and operation to joint integrity and versatility. It seems likely therefore that neither will reign supreme and both will see useful and commercial operation over a wide sphere of application.

Fig. 4.4. Micro friction welder. (Courtesy The Welding Institute.)

The area of cross-section to be joined will dictate the speed and forces necessary to effect a satisfactory joint. The range of power and speed required to cover comparatively small variations is so great that it would be mechanically undesirable to attempt to produce a machine to cope with all component sizes. For example, a five-fold increase in diameter from 12 mm to 60 mm demands an increase in applied force by a factor of 25 and a decrease in relative rotational speed of 5. Machines are thus made in a range of sizes to meet the requirements of industry. This size range extends from 1 mm diameter—micro friction welding (Fig. 4.4)—which uses typically 0·1–4 kpm and $40 - 100 \times 10^3$ rpm, to monsters which will join up to 110 mm diameter using $7\cdot6 \times 10^3$ kpm and 500 rpm (Fig. 4.5).

Fig. 4.5. 14×10^5 N, 500 rpm friction welding machine installed at the British Aluminium Company works in Invergordon. (Courtesy Black's Equipment Ltd.)

A range of medium size machines has been made to operate in the vertical position and these machines have proved to be at least as successful as the more traditional horizontal machines for high rate production of small, regularly shaped assemblies (Fig. 4.6).

Fig. 4.6. Medium capacity vertical type machine. (Courtesy Steelweld Ltd.)

Cost is examined in greater detail in Chapter 8 but it is perhaps opportune here to present an overall picture of the equipment cost range. This is shown graphically in Fig. 4.7. It should be emphasised that this presents a very arbitrary picture, since a large number of variables have been omitted for clarity.

Three variations on the basic technique of holding one part rigidly and rotating the other against it have been proposed and tested. The first of these, contrarotation, uses the principle of rotation of both parts and is useful where very high surface welding speeds are required. An alternative technique, applicable to the welding of heavy or unwieldy sections such as long pipes or tubes, is to rigidly hold two pieces and rotate a central piece between them prior to applying pressure to it. Finally, it is possible to employ a reciprocating motion as distinct from a true rotation. The latter method has not so far been found particularly attractive for any application.

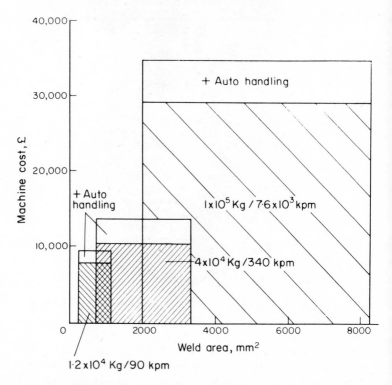

Fig. 4.7. Equipment cost range. Weld areas relate to average carbon steel requirements.[10]

5. Production Engineering

Friction welding can be compared in principle with flash butt welding, and it is in the sphere of application covered by flash butt welding that the majority of outlets have been found. The repeatability of friction welding is significantly higher than that of flash butt welding, however, and the range of materials which can be joined satisfactorily is much wider. In part, this is due to the wide parameter tolerance band which can be allowed when using friction welding. The nature of the friction-welding process is such that heat can only be imparted to the joint and its immediate surroundings. This is a further reason for the wide range of materials and combinations of materials which lend themselves so readily to being joined by friction welding.

PRE-JOINT CLEANING

Although friction welding is less sensitive to contamination than processes such as fusion welding and solid-state joining, some degree of cleaning has to be effected. However, it is necessary only to remove heavy rust and scale coating to permit adequate heat generation and thus provide conditions for bonding. Where parts have widely differing thermal masses, e.g. rod or tube to plate, special emphasis must be laid on oxide removal in order that an effective thermal situation can be created.

Lubricants and greases introduced during pre-weld machining operations or inadvertently picked up in transit can be an embarrassment. In the main, the bonding operation will move all surface contamination to the weld periphery and into the flash. It would be bad practice to rely on this happening, however, and in automatic and semi-automatic cycles where repeatability is of paramount importance, it is advisable to remove the impurities by solvent action before attempting to weld.

REPEATABILITY

In friction welding, process repeatability is undoubtedly very high indeed. This is perhaps fortunate, since non-destructive inspection after welding can be extremely difficult. Many instances of 100% repeatability have been quoted, including for example the injector assembly illustrated in Fig. 5.1.

Fig. 5.1. Injector assembly for diesel precombustion chamber. (Courtesy Production Technology Inc.)

Both major parameters, speed and applied pressure, can be controlled readily so that the process lends itself to fully automatic operation. The application to flowline work is further stimulated by the fact that a power factor of 0·85 is reached. This can be compared with 0·4–0·6 for arc and flash welding. The most difficult problems in friction welding are associated with supply and removal of components, although the development of techniques in recent years has led to the manufacture of sophisticated and reliable feed and retrieval systems. An example of a typical modern device is shown in Fig. 5.2.

Fig. 5.2. Feed system illustrating delivery of flange (right hand side) to chuck prior to friction welding or to rear axle casing. (Courtesy John Thompson Ltd.)

Operator error on fully automatic equipment is, as a consequence, low. Often the operator's responsibility is limited to merely switching on and switching off at the beginning and end of a batch or shift, and maintaining a watch on the machine output.

MAINTENANCE

High torque/high stress machines are now in common use,

and the need to sustain throughput means that maintenance assumes a particularly important role.

In the case of manually loaded machines, overall drive mechanisms including bearings, motor drive belts and equipment for applying end load must be subjected to regular inspection. When sophisticated automatic loading devices are used, special attention is needed to ensure smooth running. The rotating work-holding mechanism— usually a chuck—is probably the component part of a friction welding machine which receives most punishment and therefore requires regular inspection. Here, lack of maintenance can lead to inferior joints due to chuck jaw damage, misalignment and eccentricity.

LIMITATIONS

In friction welding, as in all material joining processes, some restriction on application is imposed by the nature

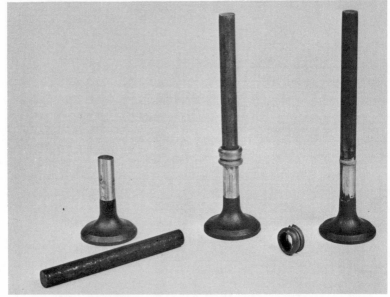

Fig. 5.3. Automobile valves. From left hand side; tensile test showing failure in stem, valve as welded, valve after hot shearing of flash. (Courtesy Production Technology Inc.)

of the materials used. Magnesium and its alloys, bronzes and combinations of titanium with other materials are particularly difficult to join. Other materials such as tungsten, cemented tungsten carbide and titanium alloys form brittle joints with steel. Despite the shortage of information currently available, it is apparent that friction welding is a more flexible technique, in respect of materials to be welded, than most other welding techniques.

Fig. 5.4. Machine used to weld and hot shear. (Courtesy Production Technology Inc.)

In the case of certain joint geometries, removal of the flash may be difficult and expensive. It can be difficult, for example, in the bore of a long tube where the weld is some distance from each end; it can be expensive if conventional machining must be used on high output rate applications. This is not to say that the flash often presents such problems. In many cases the joint can be so designed that the flash need not be removed at all: many automobiles include friction welded components with the flash left on. Automatic flash shearing after welding has become commonplace on flowline work. Figs. 5.3 and 5.4 illustrate how valve production has been simplified using a shear on the flash whilst the latter is still hot and has thus not developed full hardness.

Fig. 5.5. Machine used for phase matching of components. (Courtesy of Dr. Atsushi Hasui.)

Until recently it was necessary that one part of the assembly should be circular or at least have an axis of symmetry in rotation. Angular orientation of square or angled sections was difficult to achieve. Japanese work has now resulted in the introduction of phase matching (Figs. 5.5 and 5.6). This is still in the development stage, but there is every indication that the technique will stimulate still greater interest in the growing field of friction welding.

Fig. 5.6. Non-circular pieces welded together to produce a continuous "in phase" length. (Courtesy of Dr. Atsushi Hasui, National Research Institute for Metals, Tokyo.)

For any given machine size there is a fairly restricted parameter band which can be covered. Small parts need small machines, large joints demand large machines. No single unit is capable of accommodating the entire spectrum of joint cross-sections encountered in friction welding. Setting up a jobbing shop can thus present difficulties, as at least two machines will be necessary if a reasonable service is to be offered.

The weight of friction welding equipment restricts its mobility. Normally it will be necessary to bring the job to the machine, although there are some applications on which portable equipment can be employed (Fig. 5.7).

Workpiece shape and size dictate to a great extent whether or not friction welding can be applied. Clearly, parts which cannot be accommodated in the confines of the manipulating and rotating devices of existing machines are unlikely candidates for friction welding unless the resultant savings are such that special equipment can be justified.

39

Fig. 5.7. Portable equipment used for cable joining. (Courtesy Electricity Council Research Centre.)

WORKING PARAMETERS

A wide range of materials and workpiece sizes can be accommodated using speeds which produce a peripheral velocity of between 1200 and 1700 mm/sec. For example, 320–450 rpm is suitable for 75 mm diameter bar, 1250–1750 rpm can be used on 20 mm diameter bar. Lower velocities demand higher torques and consequently greater restraint. This introduces work-holding problems and can also lead to irregular heating. Increase in velocity produces wider heat-affected zones and increases welding time.

Friction pressure influences driving torque and power and is significant in that it controls the temperature gradient at the joint. High pressures produce rapid heating and can result in non-parallel joints. Selection of friction pressure is determined by the material to be joined. The pressure used must maintain the surfaces in intimate contact and thus prevent atmospheric contamination. At the same time it should result in uniform heating. For mild steels 30–45 N/mm^2 are used in the friction pressure cycle. Lower strength materials obviously require less pressure. High hot strength materials may demand very much higher friction pressures.

Similar considerations apply in the case of forge pressure. Here, for mild steels, 75–90 N/mm^2 can be used with confidence.

It is usual to discontinue the cycle as soon as the desired thermal conditions have been satisfied. Lengthening the cycle only results in lower productivity and greater consumption of material, although the cycle time can be used as a means of producing consistent overall lengths from components of irregular length. This is effected by allowing material to extrude axially until a desired total length has been achieved.

INSPECTION

Limitations on the application of friction welding are often due to the difficulties associated with non-destructive examination after joining. With a relatively thin interface, radiography is virtually useless. Ultrasonic techniques have developed rapidly for this purpose, but they are of little value for selective testing. They are more usefully employed on a go, no-go basis than for the location of individual defects.

Defects which do occur are usually caused either by imperfections in the parent materials or by variations in machine parameters. The former can be minimised by adequate pre-weld quality control. There is every indication that the latter almost invariably results in defects at the joint periphery which can be detected by simple visual examination or by the use of surface penetrant techniques.

With good control over materials and machine parameters, welds of the highest integrity can be produced on a highly repeatable basis. This has undoubtedly been responsible for the widespread use of friction welding even in semi-critical applications, such as motor car components.

6. Mechanical and Metallurgical Considerations

There is very little published information on the mechanical and metallurgical aspects of friction welding technique as applied to a wide range of materials. On the other hand, the friction welding of a limited range of materials has received intensive study. This latter situation has presumably arisen due to a demand from designers using friction welding for critical applications where, because of the difficulties associated with non destructive testing, it is essential to establish the mechanical and metallurgical properties and to rely upon the process repeatability to produce joints of consistently high integrity.

A balanced assessment of the mechanical properties of friction welded materials cannot yet be presented. In general terms, however, it can be said that the joint behaves in a similar manner to the brazed joint. The weld is very narrow and supported on either side by parent material. Under these conditions, simple tensile testing leads almost invariably to failure outside the joint zone. The bend test, which is more frequently applied to friction welds, can be carried out in such a manner as to stress the joint itself and thus produce results indicative of the true strength of the friction weld. A cross section of typical results is presented in Table 6(a). A more generalised illustration is presented in Table 6(b).

Additional testing, such as fatigue and rupture, has been confined to a restricted range of material joints and is an offshoot of comprehensive investigations into the more fundamental aspects of friction welding. Indeed, information is so scant that separate examination of how friction welded joints behave generally under conditions

Table 6(a). Cross-section of some typical mechanical test results on friction welded joints

MATERIALS COMBINATION	TENSILE (MN/m²)	BEND (degrees)	ELONGA-TION (%)	IMPACT (kpm)
EN3A–EN2A	440	>100	25	—
AISI.4140–AISI.304	670	>180	44	—
2024–T4 aluminium (aged at ambient temp. for 72 hr. after welding)	470	80	>30	0·7
7075–T6 aluminium (aged at ambient temp. for 48 hr. after welding)	450	48	20	0·7

of fatigue, etc., cannot be justified: relevant information must be extracted from each investigation in turn.

Torsional fatigue tests on joints in SAE1035 boron treated steel revealed that up to 69 MN/m² properties of the welds were within the scatterband of the parent material strength. Considerable scatter occurred above 100 MN/m² and reliable results would necessitate further work.

Valve head and stem combinations, stellite 6 and SAE8645 respectively, have been tested in push–pull fatigue. Failures occurred in the stellite away from the joint up to 480 MN/m² and $2·5 \times 10^{-6}$ cycles.

Rupture behaviour on joints between cast GMR235 (a high chromium–nickel alloy) and A286 used for blade hub/blade ring fabrication indicates encouraging results at 650°C. At 470 MN/m², for example, 107 hours was required for failure. Further information on this type of application is very limited but both the General Electric Co. and Pratt and Whitney are exploring the possible use of friction welding on advanced gas turbine engines.

Table 6(b). Tabulated data for materials on which attempts have been made to effect joints by friction

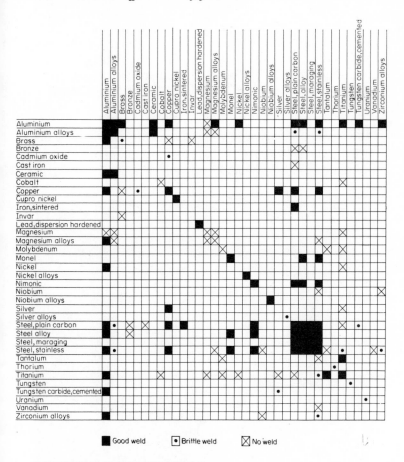

■ Good weld   ● Brittle weld   ✕ No weld

DIFFICULT MATERIALS

One of the greatest assets of friction welding is that it can be used successfully with materials which are either impossible or at best extremely difficult to join by other processes.

Self-lubricating alloys can present problems due to difficulties in creating frictional heat. For example, free machining grades of steels can be welded but highly resulphurised and leaded types do not attain full strength although metallurgically the joints are of high integrity. Widely dissimilar materials such as aluminium and steel give rise to brittle joints when welded by alternative methods. Friction welds in these metals exhibit superior properties. Indeed, as will be seen in Chapter 7, an aluminium/steel joint of 200 mm diameter is used regularly by the aluminium smelting industry. Joints between aluminium and copper also exhibit high strength.

Where two materials of differing ductility are joined the softer one will suffer the greater loss of material, and allowance must be made for this in assessing lengths (Fig. 6.1).

Fig. 6.1. Sound friction welded joint in mild steel. (Courtesy the Welding Institute.)

CONTAMINATION

The minimal pre-joint cleaning requirements were discussed in Chapter 5. With most materials a dry surface in the "as sawn" condition can be welded directly. Other materials are more sensitive to surface preparation but, in general, impurities are extruded into the flash leaving metallurgically sound joints (Fig. 6.2). With any given material, friction welding demands much less attention to cleanliness than other methods of welding.

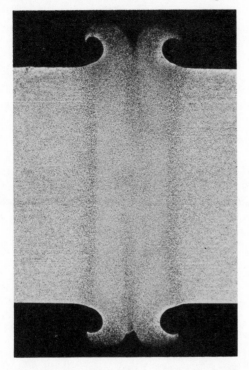

Fig. 6.2. Joint between aluminium and steel showing massive upset in the softer aluminium. (Courtesy the Welding Institute.)

HEAT INPUT

The inherent nature of the friction welding process means that only a small zone is thermally disturbed. In this respect, friction welding is comparable with electron beam welding. Because of the small heat input, however, cooling rates after joining can be high; a large temperature differential exists over a small axial length. Hardenable steels thus tend to exhibit hard heat-affected zones. Even

47

in mild steel joints peak hardnesses of nearly 500 VPN have been recorded.

Heat treatment after welding is often effective in reducing hardness, but care is needed to ensure that other properties are not impaired. Overtempering, gross diffusion and carbon migration may occur if heat treatment is used indiscriminately.

Slowing down the overall friction welding cycle can be most effective in reducing joint hardness, since a greater volume of metal will experience a thermal cycle. As a consequence larger heat-affected zones must be tolerated. Production times will also be prolonged.

Published results and information from a wide cross section of industry indicate that friction welding is capable of producing joints of high integrity. The process is metallurgically and mechanically competitive with alternative welding techniques. Friction welding also offers a means of welding many materials which are virtually unweldable by other joining methods.

7. Design and Application

Friction welding may be used effectively on a comparatively restricted range of component shapes. The selection of joint design is thus simplified. Emphasis is usually placed on the positioning of the joint in the area of least stress in applications where strength in service is essential. A further consideration is the significance of the flash, which may have to be removed for mechanical or aesthetic reasons.

Circular cross sections provide the ideal surfaces for joining, but some deviation is acceptable. Until comparatively recently, it has not been possible to effect alignment of irregular cross sections, but his now appears to have been overcome (see Chapter 5).

The surfaces to be joined require little or no preparation. Sawn or sheared edges are often prejoined without any other treatment. Forgings can be welded in the "as received" condition if forge scale is removed.

Due to the high torque loading on components, design must be such as to allow for rigid clamping in order that slippage during welding does not occur. In general, parts should be gripped on a diameter equal to or greater than the weld outside diameter and as close to the weld plane as the geometry will allow.

Present equipment allows for the welding of size ranges from under 0·25 mm to over 600 mm diameter and up to 6 m long. Production applications should have a wall thickness greater than 10% of the diameter to preclude any possibility of buckling or collapse. Special holding devices will be essential to accommodate thinner components.

Reduction in overall length inevitably occurs during friction welding. Typically, in 25 mm diameter solid mild steel bar about 8 mm loss will be sustained. Tolerance on final length, however, can be controlled readily to ± 0.25 mm whilst still accommodating variations in component lengths of ± 1.5 mm. This is possible since machine parameters can be set by limit switches which control the burn-off phase to produce a predetermined final length.

Concentricity can be held to ± 0.25 mm on 25 mm diameter bar. With increasing cross sections eccentricity becomes greater.

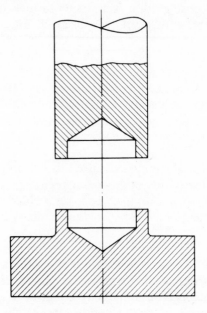

Fig. 7.1(a). Centre relief designed into bar joint to decrease cross sectional area at weld interface.

In Chapter 5 it was explained why a single machine could not cater for a wide size range of joints. By intelligent design, however, the versatility of any given machine can be extended to accommodate larger joints than it was intended for. For example, it will often be mechanically acceptable to provide a centre relief in a bar (Fig. 7.1(a)). This lowers the cross sectional area to be welded and therefore the pressure, thrust and torque requirements. It should be noted that internal flash removal will not be possible with this design unless the centre relief can be extended down the axis of one component (Fig. 7.1(b)).

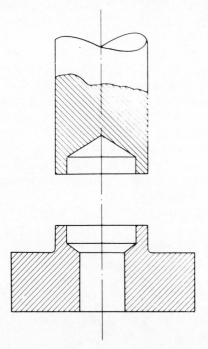

Fig. 7.1(b). Design which allows access for internal flash removal.

INDUSTRIAL APPLICATIONS

Bearing in mind the short time that friction welding has been in existence as a commercial tool it has found an astonishing variety of applications. The widespread acceptance of friction welding, particularly in the mass production industries, is a reflection of the confidence placed on the repeatability when machine parameters have been established by development to give optimum joints.

The following examples have been selected to illustrate current industrial uses of friction welding.

Internal Combustion Engine Valves (Fig. 7.2)

During recent years, increasing engine speeds have created a demand for valve materials of improved properties. In general, materials were required to provide valve stems of greater wear resistance, and heads of greater heat and corrosion resistance. One way of meeting this requirement was to construct a bimetallic valve using nickel based alloys for the head and a hardenable steel for the stem. Initially, flash butt welding was employed but with reject rates in the 15% region the process was never considered really acceptable. The adoption of friction welding reduced the rejection rate virtually to nil, providing an example of the way in which friction welding can solve a technical problem and at the same time effect a significant cost saving.

Fig. 7.2. Internal combustion engine valve together with selected examples of other automobile components. (Courtesy Steelweld Ltd.)

In Germany, flash butt welding has been replaced by friction welding for the manufacture of Nimonic-headed valves. The production rate is in excess of 13 per minute and concentricity has been improved by 300%.

The Renault factory in France produces 90,000 valves per month at a rate of 8 per minute, flash being removed automatically during the cycle.

Rotating Assemblies (Fig. 7.3)
Significant savings in raw materials have followed the introduction of friction welding in the manufacture of impellers, superchargers and similar products where the geometry is essentially a disc welded axially to a shaft.

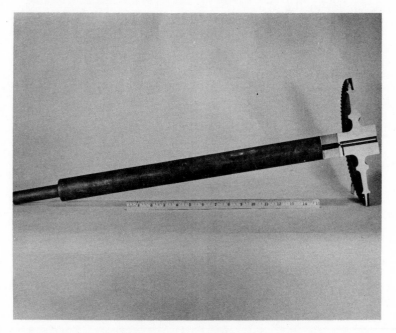

Fig. 7.3. Rotor disc friction welded to shaft after removal of outer flash. The weld can be seen clearly in the section taken. (Courtesy Production Technology Inc.)

The Garret Corporation of America produces 6000 turbo-chargers a month, using steel shafts and nickel alloy wheels. Operating at 135,000 rpm and 900°C the assembly has proved entirely satisfactory.

In Japan, heat-resistant cast wheels are friction welded to steel axles for exhaust gas turbines in superchargers.

Road Vehicles
In addition to its use in valve manufacture, friction welding is also used extensively in other components of road vehicles.

In the U.K., the Ford Motor Company have made over 2 million welds between steel tubes and forgings at a rate of over 100 per hour for rear axle housings. Chrysler and British Leyland also manufacture this component by similar methods.

In the U.S. the Frauhof Corporation uses friction welding in the production of S cams used in the actuation of trailer brakes (Fig. 7.4). The cams were made previously from solid forgings.

Fig. 7.4. Friction welded trailer brake cam. (Courtesy Production Technology Inc.)

Printing Industry
Miehle-Goss Dexter Inc., a leading U.S. manufacturer of newspaper presses, has adopted friction welding in the production of heavy rollers (Fig. 7.5). Made in a low carbon steel, the rollers are 2 m long and 150 mm diameter. A 40 mm diameter journal is welded to each end. After two years in operation there have been no incidences of weld failure.

Fig. 7.5. Heavy hollow roller, 150 mm diameter, on the right hand side welded to end cap of the same diameter and then to the journal on the left hand side. (Courtesy Production Technology Inc.)

Fig. 7.6. Aluminium hanger being removed from friction welding machine. The joint is where the transition from rectangular to circular section occurs. (Courtesy Black's Equipment Ltd.)

55

Aluminium Hangers (Fig. 7.6)

Probably the largest section weld currently made by friction welding is between aluminium and steel. A $(14 \times 10^{-5}$ N) machine is used by British Aluminium in producing anode hanger assemblies for use in smelting baths. The aluminium section $(110 \times 120$ mm), is up to 3 m long and is butt welded onto a steel bar of more than 200 mm diameter. An estimated 50% cost saving has been made over the previously used mechanical method of joining.

8. The Cost of Friction Welding

The fact that friction welding has found so many applications in mass-production industries provides an indication of its effectiveness. In the production of components for motor cars, for example, the adoption of friction welding has resulted in nett savings in excess of 15%.

Smaller organisations operating subcontract friction welding services on small batch runs are offering material savings, wide design latitude and high speed production.

Friction welding is inherently efficient and requires less power than the nearest alternative—flash butt welding— to produce a similar end result. The process lends itself to fully automatic operation, and unskilled labour can be employed.

The machining of surfaces to be joined, often a costly operation, is commonly necessary in the case of fusion welding techniques; it can be eliminated, however, by using friction welding. Moreover, friction welding does not require filler metals, gas or flux, and this too can result in significant economies.

The cost of friction welding machines, the range of cross sections which can be welded and the output rates are summarised in Fig. 4.7 and Table 8(a). These indicate the wide range of joint types and sizes that can be accommodated by currently available equipment.

Manufacturers are commonly reluctant to release detailed information on costs, but a number of friction welding applications have been analysed in depth. The following examples have been selected to illustrate the economies that can be made by using friction welding.

Table 8(a). Review of component sizes, output rates and nominal prices

BAR DIA. (mm)	FORGE LOAD CAPACITY (kg)	OUTPUT PER HOUR	PRICE RANGE (£1,000)
Up to 10	5,000	500	10
10–35	15,000	250–120	10–15
30–65	50,000	120–60	14–20
50–100	100,000	60–20	More than 30

PISTON RODS (Fig. 8.1)

Hydraulic piston rods of various lengths are manufactured by Caterpillar Corporation of Illinois, U.S.A. Prior to the application of friction welding a single forging was used, and an inventory of 155 sizes was necessary to guarantee continuity of supplies. Following extensive mechanical testing on the joints it was established that fatigue properties and reliability of a friction welded assembly was as good as the single piece forging. A nett annual saving in direct costs exceeding £50,000 ($125,000) has been realised and the inventory reduced from 155 forgings in various lengths and sizes to only 26 eye forgings. Standard bar stock is used for the shaft and only requires to be cut to length prior to welding.

Fig. 8.1. Eye forging as welded and with sectional view of finished part. (Courtesy Production Technology Inc.)

FLASH BUTT WELDING SUPERSEDED

Larger sizes (16 mm diameter and upwards) of high speed steel twist drills have for some time been manufactured by joining a high speed steel flute to a carbon steel shank using flash butt welding. The introduction of friction welding by Samuel Osborn Ltd. of Sheffield resulted in a reduction of metal loss during the joining cycle and improved weld integrity. Straightening after joining was no longer necessary.

Considering only the reduction in materials used, a saving of £7,000 ($17,500) per annum was recorded on a machine producing 110 drills per hour (Fig. 8.2).

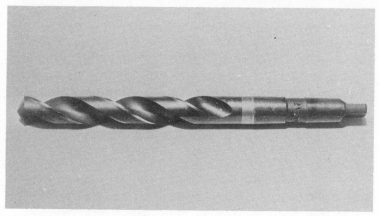

Fig. 8.2. Friction welded and finished twist drill. (Courtesy Production Technology Inc.)

MATERIAL SAVINGS

Spindles made by Caterpillar Corporation (Fig. 8.3) are produced in limited numbers and do not justify forging. Machining from solid bar would be very expensive. The friction welding of bar stock of differing diameters shows a 52% material saving, providing a joint which is as strong as the solid metal.

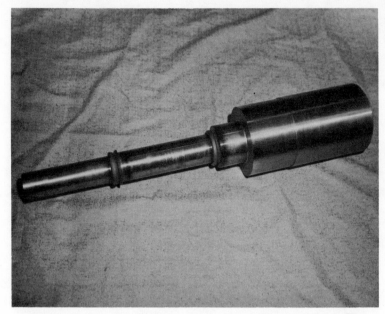

Fig. 8.3. 140 mm diameter heavy bar welded on to smaller ameter bars to effect material cost savings. (Courtesy Production Technology Inc.)

MACHINING PROBLEM SOLVED

The component illustrated in Fig. 8.4 has multiple applications as an airborne device and is manufactured for the aerospace and defence group of Honeywell Inc., U.S.A. The 11 mm diameter tubular section was previously machined from solid stock and involved an expensive deep hole drilling operation. Using friction welding, screw machine end fittings are now joined to both ends of a tube which makes up the long centre section. The friction welded unit is functionally interchangeable with the original part.

Designing the joint welding presented a problem, since flash could not be tolerated beyond a designated axial distance inside the tube. The problem was solved by using an internal pin during the joining cycle to control the flow of extruded material. This can be seen clearly in Fig. 8.4. Production costs have been cut by 50% and a 30% reduction in weight has been achieved.

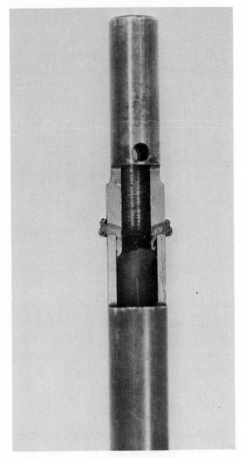

Fig. 8.4. Section through tubular joint showing control which has been exercised over inside flash by the use of a removable internal pin. (Courtesy Production Technology Inc.)

ELIMINATION OF BRAZING

Squibb Pitzer Inc. of Dallas, U.S.A., have effected major savings in the manufacture of bomb fusewells (Fig. 8.5).

Originally the joint at the base was made using a silver-based braze metal and involved 14 staff working a 24-hour day preparing assemblies for furnaces. Despite careful quality control an embarrassingly high reject rate was unavoidable. The joint is now friction welded at a rate of 350 per hour, two operators working alternately on a nine-hour shift being adequate to supervise the process. The reject rate is negligible.

Fig. 8.5. Cold drawn thin-walled component 75 mm diameter friction welded to solid base. (Courtesy Production Technology Inc.)

ALTERNATIVE TO FUSION WELDING

High speed welding of track roller rims, shown in section in Fig. 8.6 has been rendered possible by friction welding a joint previously made by automatic submerged arc welding. Elimination of preheating and simplified joint design have led to immediate savings of 17% at the Illinois plant of the Caterpillar Corporation.

Fig. 8.6. Track roller rim fabricated by friction welding. (Courtesy Production Technology Inc.)

SMALL BATCH SUBCONTRACT WELDING

Friction welding has proved particularly effective in applications involving batch sizes of a million parts and more; nevertheless, there is profit to be made on small batch runs. Gardena Inc. of California, for example, offer a friction welding service which has resulted in 30–35% savings to the customer in production costs over a range of applications.

9. The Future

Like all new processes, friction welding is undergoing continuous technical improvement. Innovations in work handling coupled with further experience in the mechanical and metallurgical aspects of joints will probably be accompanied by more widespread application of the technique, particularly in areas where dissimilar metals need to be joined.

Although stress has been laid on the use of fully automatic machines, a relatively small proportion of equipment in current use is in fact mechanised to this extent. Since friction welding lends itself so readily to complete automation and virtual displacement of the operator there is every indication that a swing in this direction is inevitable. This is particularly so on flow line work where there is pressure to eliminate errors due to operator control.

Friction welding came into commercial use only during the late 1960s, but the technique is now widespread throughout the industrial world. Russia and Japan each has some three to four hundred welding machines in operation (1972); there are over two hundred units in Europe and a similar number in the U.S. The rate of capital investment shows no sign of slowing down. Indeed, as the true potential of the process becomes more widely appreciated there is little doubt that the rate will increase.

REFERENCES

1. Vill, V. I., *Friction Welding of Metals* (A.W.S., New York, February 1962).
2. Kragel'skiy, I. V. and Vinogradova, I. E., *Friction Coefficients* (Mashgiz, 1955).
3. Hazlett, T. H., "Fundamentals of Friction Welding", *Metals Eng. Quarterly*, (A.S.M., February 1967).
4. Bowden, F. B. and Tabor, D., *The Friction and Lubrication of Solids*, Part I (Oxford University Press, 1954).
5. Goddard, J. and Wilman, M., "A Theory of Friction and Wear During the Abrasion of Metals", *Wear*, vol. 5 (1962).
6. Report of U.S. Welding Delegation on visit to U.S.S.R. (A.W.S., New York, 1962).
7. Rykalin, N. N., Pugin, A. I. and Vasil'eva, V. A., "The Heating and Cooling of Rods Butt Welded by the Friction Process", *ibid* (October 1959).
8. Weiss, H. D. and Hazlett, T. H., "The Role of Material Properties and Interface Temperatures in Friction Welding Dissimilar Metals", *A.S.M.E.*, 66–Met–8 (February 1966).
9. Bland, J., "Friction Welding", *A.S.M.E.*, 67–DE–25 (March 1967).
10. Ellis, C. R. G., "Friction Welding—What it is and how it works", *Metal Construction* (May 1970).

INDEX

M & B MONOGRAPHS

M & B TECHNICAL LIBRARY